Simple Machines
LEVERS

David Glover

Heinemann Library
Chicago, Illinois

© 1997, 2006 Heinemann Library
a division of Reed Elsevier Inc.
Chicago, Illinois

Customer Service 888-454-2279
Visit our website at www.heinemannraintree.com

Designed by Victoria Bevan and Q2A Creative
Illustrated by Barry Atkinson (p. 15), Douglas Hall (pp. 18, 19), Tony Kenyon (pp. 5, 9),
and Ray Straw (p. 12) Printed in China by WKT Company Ltd.

10 09 08 07 06
10 9 8 7 6 5 4 3 2 1

New edition ISBN: 1-4034-8563-1 (hardback)
              1-4034-8592-5 (paperback)

The Library of Congress has cataloged the first edition as follows:
Glover, David, 1953 Sept. 4-
    Levers / David Glover.
        p. cm. -- (Simple Machines)
    Includes index.
    Summary: Introduces the principles of levers as simple machines, using examples from
everyday life.
        ISBN 1-57572-080-9 (lib. bdg.)
        1. Levers - Juvenile literature. [1. Levers] I. Title. II. Series: Glover, David, 1953
Sept. 4 – Simple Machines.
TJ147.G65 1997
621.8' 11—dc20

                                                                                    96-15816
                                                                                         CIP
                                                                                          AC

Acknowledgments
The author and publishers are grateful to the following for permission to reproduce
photographs: Trevor Clifford pp. 3, 4, 6, 7, 8, 9, 10, 11, 12, 14, 15, 21, 22, 23; Collections/Keith
Pritchard p. 16; Spectrum Colour Library p18; Tony Stone Images p. 17; Zefa p. 13.

Cover reproduced with permission of Alamy.

The publishers would like to thank Angela Royston for her assistance in the preparation
of this edition.

Every effort has been made to contact copyright holders of any material reproduced in this book.
Any omissions will be rectified in subsequent printings if notice is given to the publisher.

The paper used to print this book comes from sustainable sources.

Contents

Some words are shown in bold, **like this**. You can find the definitions for these words in the glossary.

What Are Levers?

A lever is a rod or bar that makes things move. A seesaw is a lever. It has a **pivot** in the middle where it turns. If you push down on one end of a seesaw, the other end goes up.

A light person can balance a heavy person on a seesaw. The heavier person must sit closer to the pivot.

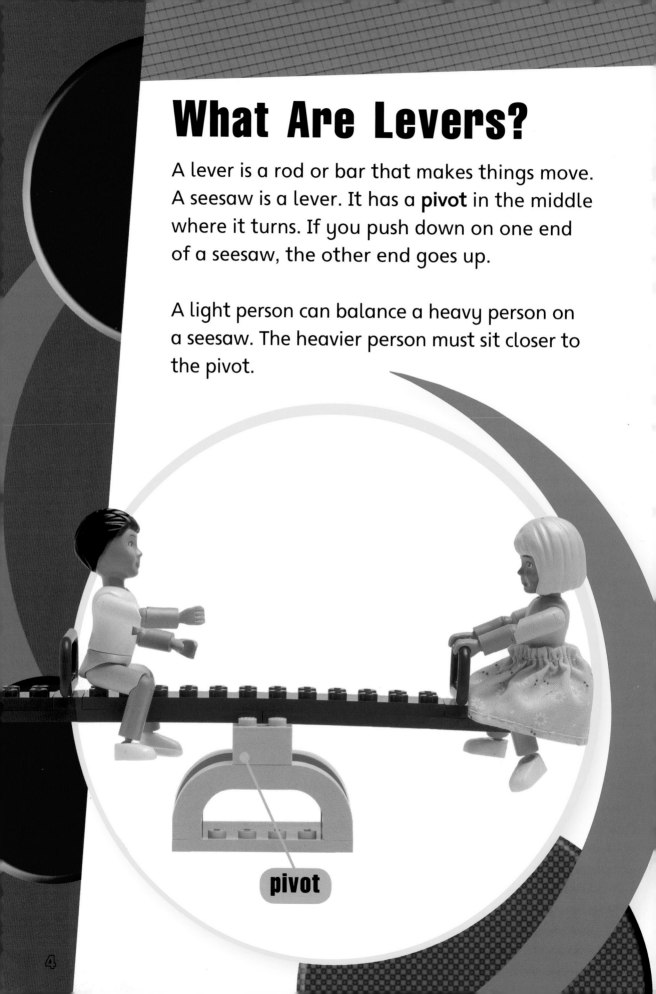

pivot

This plank is a lever.
It pivots at one end.
If the **load** is near the
pivot, you can use a lever
to lift a heavy load with
a small **effort**.

Levers do many
different jobs inside
machines. They change
pushes into pulls and
they balance weights.
Levers also move big
loads with little effort.

5

Openers

Have you ever opened a paint can with the handle of a spoon? The lid fits too tightly to open with your fingers. But when you use the spoon to open the can, the lid comes off easily.

The spoon handle is the lever. Your hand makes the **effort**. The stiff lid is the **load**. The place where the handle rests on the edge of the can is the **pivot**.

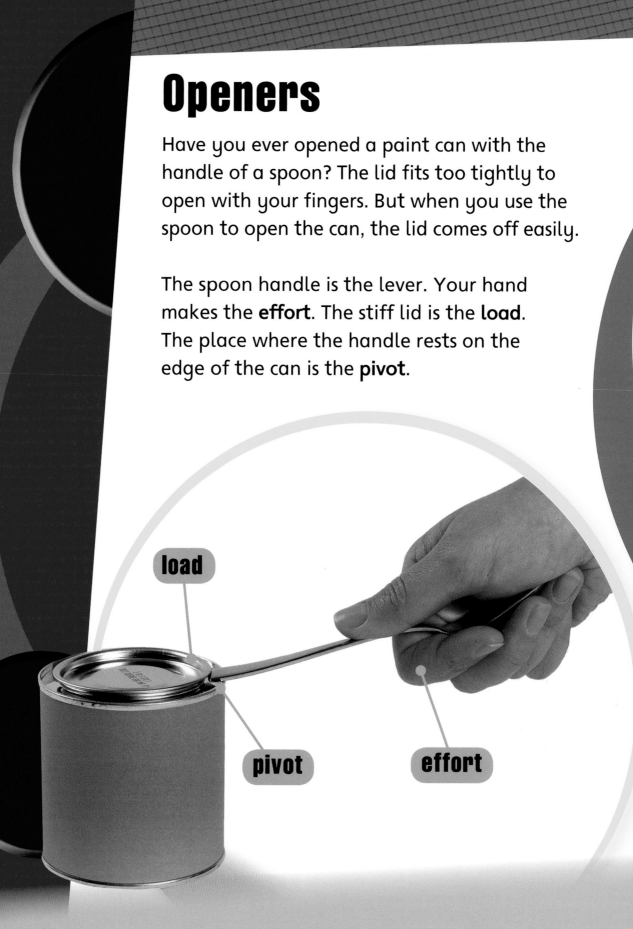

load

pivot

effort

Magnify your strength!

Levers can magnify your strength. They increase the effect of your strength. With a lever, you can push or pull much harder than you can with your bare hands. This is why levers are so useful.

A bottle opener is also a lever. It pivots on the bottle cap. Your effort as you lift the end of the opener lifts the bottle cap off.

7

Barrows

A barrow works as a lever to lift heavy **loads**. A gardener's wheelbarrow **pivots** around the wheel. The gardener's **effort** on the handles is farther away from the pivot than the load. The gardener can lift more in a barrow than in her bare arms.

load

pivot

effort

This man's barrow has long handles. It lifts heavy suitcases off the ground. When the suitcases rest over the wheels, the man can balance them with little effort.

FACT FILE **Moving the world!**

In ancient Greece, the scientist Archimedes knew that levers could magnify strength. He said, "Give me a long enough lever and I will move the world!"

Tools

You are not strong enough to pull out a nail with your fingers. But you can pull it out with a **claw hammer**. When you pull on the handle, the hammer **pivots** on its head. The claws grip the nail and drag it out of the wood.

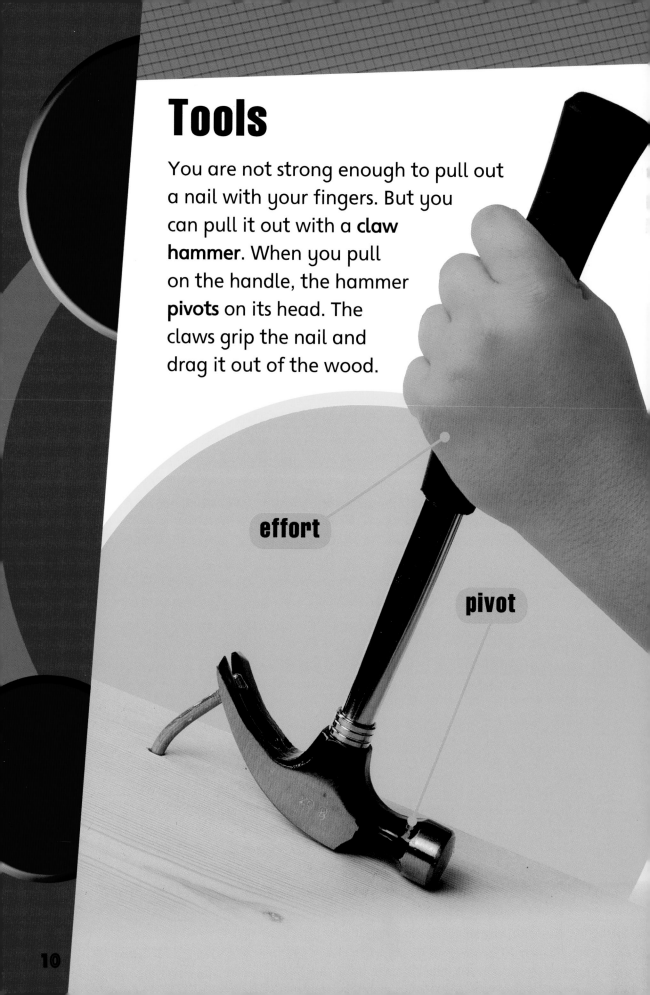

effort

pivot

liers are a pair of levers. The jaws
re close to the pivot. Pliers can
rip with great strength because
hey have long handles.

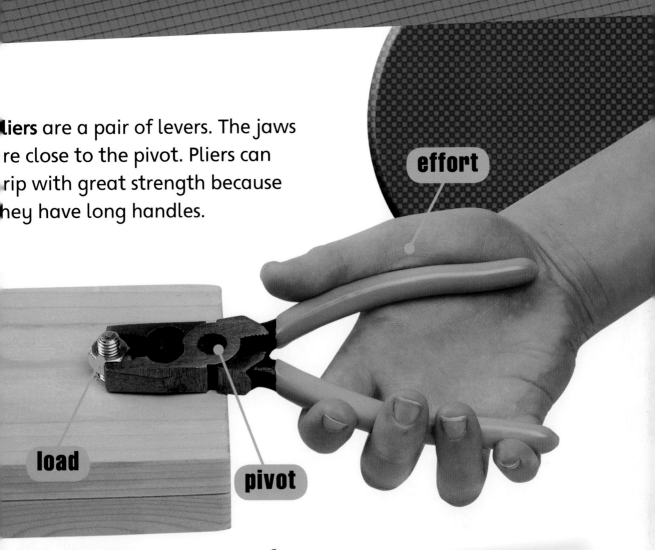

effort

load

pivot

FACT FILE Jaw power!

Your jaws are a pair of
levers. They work like
pliers. Your back teeth
are closer to the hinge of
your jaws than your front
teeth, so they can bite
harder. That is why it is
easier to crunch a carrot
with your back teeth.

11

Crackers and Cutters

Nutcrackers are a pair of levers with a **pivot** at one end. You can prove that they **magnify** your strength. First, try cracking a hard nut between your fingers. Unless you are Superman, you will not be able to. Now, use the nutcrackers—it will make a big difference!

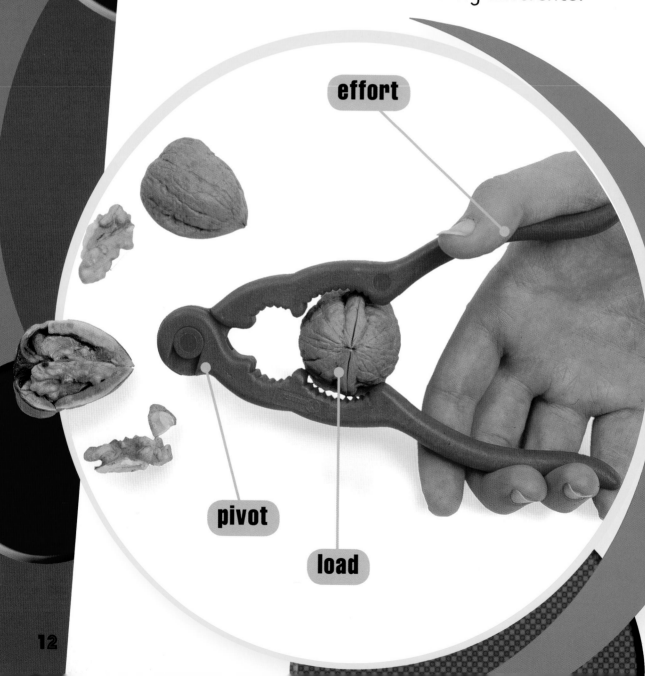

effort

pivot

load

The long handles on these cutters magnify the **effort** from your arms. They can snip through thick branches. With cutters like this, you can even cut through metal bars.

Cracking chimps!

Chimpanzees can crack nuts with stones. However, the human being is the only animal to use levers.

Balances

This toy balance is a lever, just like a seesaw. You can experiment to see how far from the **pivot** you have to put different weights in order to make them balance.

Can you balance two weights on one side with one weight on the other side? The weight on its own must be twice as far from the pivot to make it work.

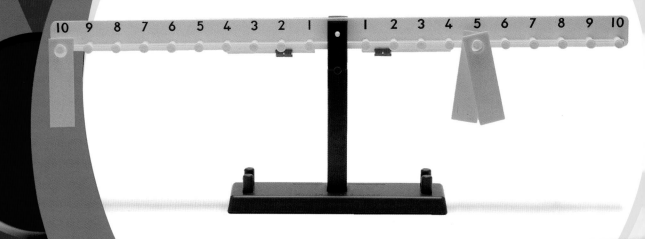

These kitchen scales use a lever to find when the weights are balanced. If you want to weigh some apples, you put the apples in one pan. Next, you add the weights to the other pan. If the apples weigh more than the weights, their pan stays down. If the weights weigh more, the apples go up. If the weights are equal, the pans are exactly balanced.

A good balance is very accurate. The weight of one feather is enough to "tip the balance" one way or the other.

feather

Bridges

This bridge has to be lifted out of the way to let a boat pass by. It is fixed to a lever. When one end of the lever arm is pulled down, the other end pulls the bridge up into the air.

There is a heavy weight on one end of the lever arm. This weight balances most of the weight of the bridge. A person can lift the bridge with just a little extra **effort**.

Tower Bridge in London, England, is a lever bridge. Its decks weigh about 1,100 tons (1,000 tonnes). They are balanced by huge weights. When a ship needs to pass under the bridge, electric **motors** lift up the decks.

Rods and Oars

A fisherman uses levers, because a fishing rod is a lever. He uses one hand as a **pivot** and the other one for the **effort**. Small movements of his hands are **magnified** by the long rod. When the fisherman gets a bite, he can move his rod very quickly.

load

effort

pivot

The oars on a rowing boat are levers, too. You pull on the oars to move the boat through the water. The oars pivot in the **oarlocks**.

effort

load

oarlock (pivot)

Brakes

Could you stop a heavy carriage like this with one hand? The answer is yes, if you use a brake lever. The lever pushes a wooden block onto the wheel. **Friction** slows the wheel down. Friction is the force that stops things from sliding easily.

brake lever

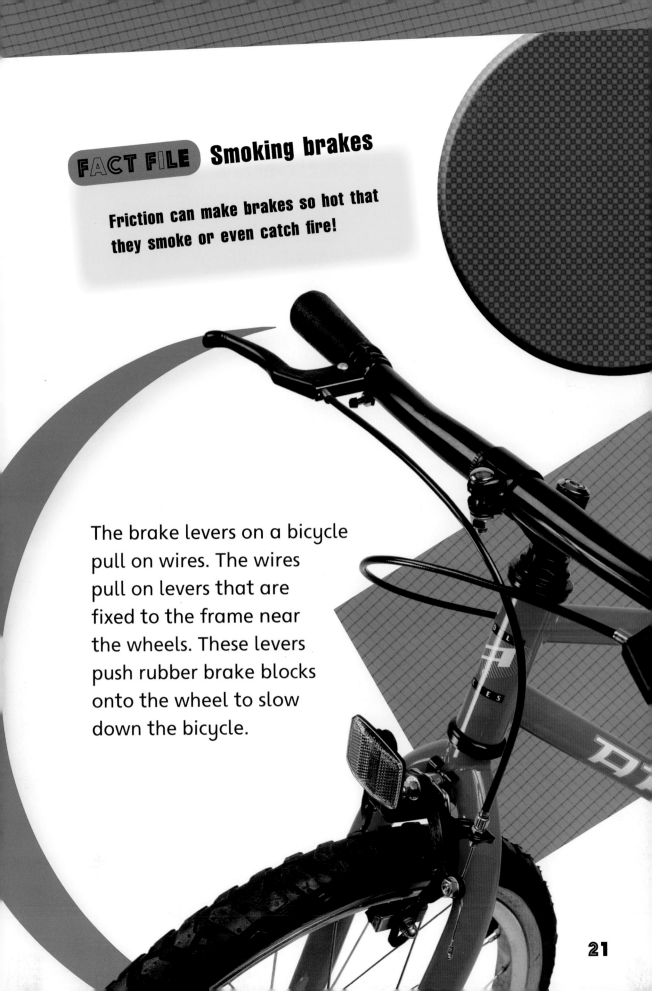

Smoking brakes

Friction can make brakes so hot that they smoke or even catch fire!

The brake levers on a bicycle pull on wires. The wires pull on levers that are fixed to the frame near the wheels. These levers push rubber brake blocks onto the wheel to slow down the bicycle.

Activities

Which is the best lever?

1. Balance a long ruler over a **pivot**, such as a block of wood.
2. Rest another piece of wood on one end.
3. Push down on the other end of the ruler to lift up the wood.
4. Repeat this, using a standard 12-inch (30-cm) ruler. Then, use a short 6-inch (15-cm) ruler.

Which ruler makes the best lever?

Make it balance

1. Make a balanced seesaw using a ruler and a pivot.
2. Put a weight on one end and two weights on the other end. What happens?
3. Move the heavy weights toward the pivot until the seesaw balances again.
4. Repeat this using different numbers of weights on the heavier end.

This is explained on pages 4 and 14.

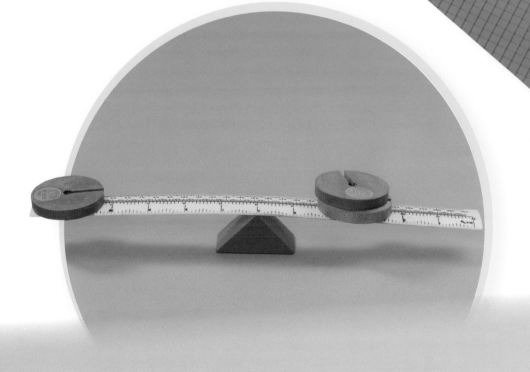

Glossary

claw hammer   hammer with two claws on the back of the head

effort   push or pull you use to move something

friction   drag or force that stops one thing from sliding over another smoothly

load   something, usually heavy, that you are trying to move

magnify   make bigger

motor   machine that uses electricity or fuel (such as gasoline or coal) to make things move

oarlocks   grips on the sides of a boat that hold the oars in place as you row

pivot   place around which a lever turns

pliers   gripping tool with jaws that you can close by squeezing a pair of handles

Index